US MILITARY BRANCHES

THE US SPACE FORCE IN ACTION

Percy Leed

Lerner Publications ◆ Minneapolis

Copyright © 2023 by Lerner Publishing Group, Inc.

All rights reserved. International copyright secured. No part of this book may be reproduced, stored in a retrieval system, or transmitted in any form or by any means—electronic, mechanical, photocopying, recording, or otherwise—without the prior written permission of Lerner Publishing Group, Inc., except for the inclusion of brief quotations in an acknowledged review.

Lerner Publications Company
An imprint of Lerner Publishing Group, Inc.
241 First Avenue North
Minneapolis, MN 55401 USA

For reading levels and more information, look up this title at www.lernerbooks.com.

Main body text set in ITC Franklin Gothic Std.
Typeface provided by Adobe Systems.

Editor: Lauren Foley **Designer:** Mary Ross

Library of Congress Cataloging-in-Publication Data

Names: Leed, Percy, 1968– author.
Title: The US Space Force in action / Percy Leed.
Description: Minneapolis: Lerner Publications, [2023] | Series: US military branches (Updog books) | Includes bibliographical references and index. | Audience: Ages 8–11 | Audience: Grades 2–3 | Summary: "The US Space Force helps protect and maintain US satellites. Readers will explore what Space Force guardians do and the unique technology and vehicles they use"—Provided by publisher.
Identifiers: LCCN 2021047496 (print) | LCCN 2021047497 (ebook) | ISBN 9781728458328 (library binding) | ISBN 9781728463636 (paperback) | ISBN 9781728462547 (ebook)
Subjects: LCSH: United States. Air Force Space Command—Juvenile literature. | Satellites—United States—Juvenile literature.
Classification: LCC UG1523 .L44 2022 (print) | LCC UG1523 (ebook) | DDC 358/.800973—dc23/eng/20211006
LC record available at https://lccn.loc.gov/2021047496
LC ebook record available at https://lccn.loc.gov/2021047497

Manufactured in the United States of America
1-50861-50198-1/27/2022

TABLE OF CONTENTS

About the Force   4

Vehicle Close-Up   12

On Base   14

Studying Space   18

Military Mission   21

Glossary   22

Check It Out!   23

Index   24

ABOUT THE FORCE

Over 22,000 miles (35,406 km) away, satellites go around Earth in space.

satellite: a spacecraft that moves around a sun, moon, or planet

On Earth, a US Space Force member uses a computer to protect its satellites from attacks.

Space Force keeps its satellites safe.

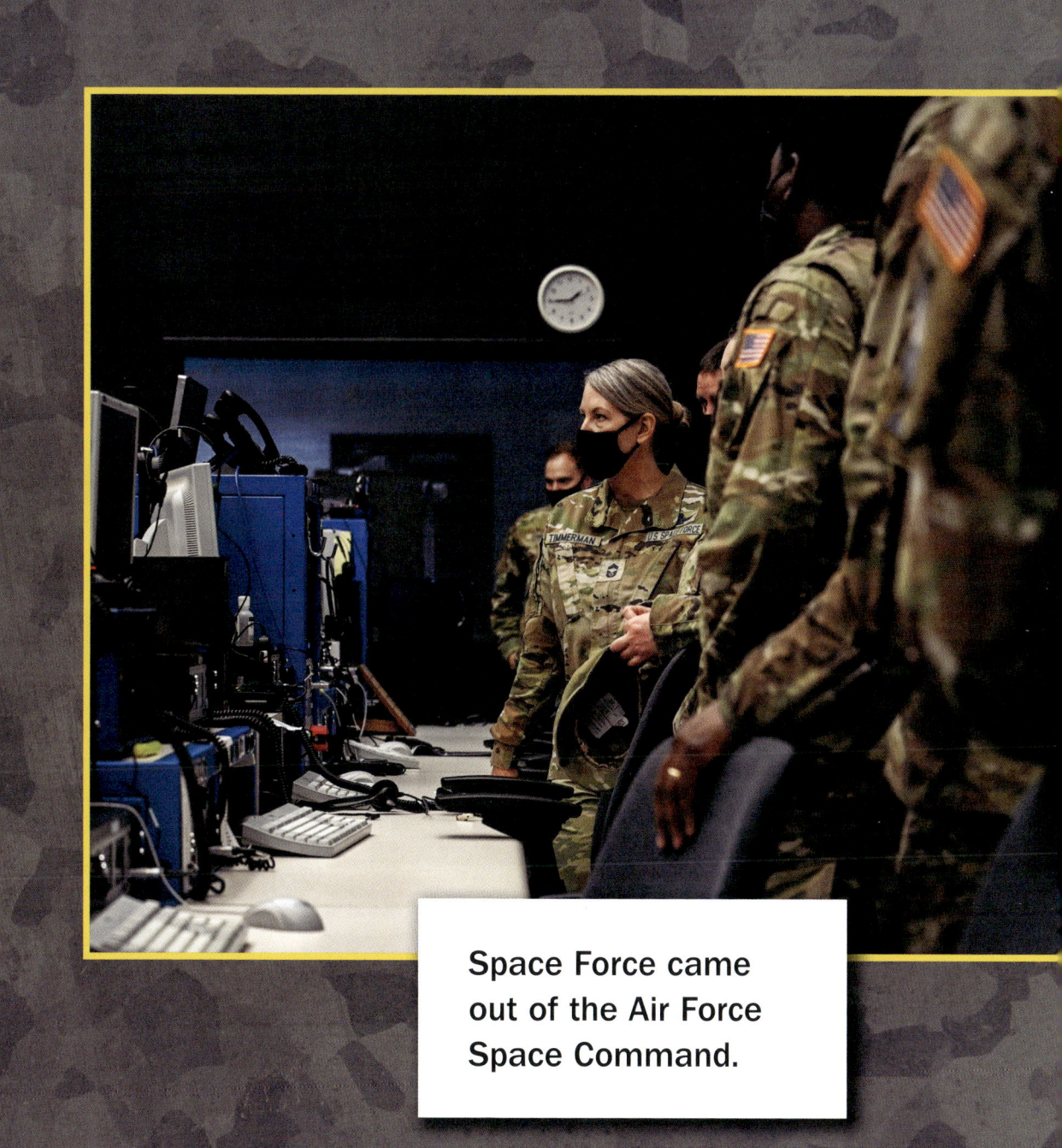

Space Force came out of the Air Force Space Command.

Space Command started in 1982 and launched satellites.

The satellites helped create GPS and find launched missiles.

GPS: a system that uses satellites to tell you where you are or how to go somewhere

missile: a rocket that explodes

9

In 2019, Space Force became its own military branch.

About sixteen thousand people became guardians.

guardian: someone who works for Space Force

11

VEHICLE CLOSE-UP

Space Force launched the Northrop Grumman Pegasus XL rocket in 2021. It is small but fast. It can quickly reach many different orbits.

nozzle

fins

orbit: the path one object takes to go around another object

nose

UP NEXT!
Inside the Force.

13

ON BASE

Space bases are located around the US.

The two biggest bases launch satellites.

There are a lot of jobs on base.

Some people fix cables so communication systems work. Others retrieve spacecraft.

UP NEXT!
Blastoff.

17

STUDYING SPACE

NASA and Space Force sometimes work together.

They want to bring astronauts to the moon again.

Space Force keeps its eye on the sky. They want to learn more about space.

MILITARY MISSION

SPACE FORCE WANTS TO IMPROVE GPS. WHAT SHOULD THEY DO?

A. Send astronauts to space

B. Launch a satellite

C. Talk to the air force

Answer: B

GLOSSARY

GPS: a system that uses satellites to tell you where you are or how to go somewhere

guardian: someone who works for Space Force

missile: a rocket that explodes

orbit: the path one object takes to go around another object

satellite: a spacecraft that moves around a sun, moon, or planet

CHECK IT OUT!

Bard, Jonathan. *How a Satellite Is Built*. New York: Gareth Stevens, 2021.

Finan, Catherine C. *Space Jobs*. Minneapolis: Bearport, 2022.

Kiddle: US Space Force
https://kids.kiddle.co/United_States_Space_Force

Leed, Percy. *The US Air Force in Action*. Minneapolis: Lerner Publications, 2023.

NASA: How Does GPS Work?
https://spaceplace.nasa.gov/gps-pizza/en/

NASA: How Do We Launch Things into Space?
https://spaceplace.nasa.gov/launching-into-space/en/

INDEX

Air Force Space Command, 7–8

base, 14–16

GPS, 9

missile, 9

NASA, 18

satellite, 4–6, 8–9, 15

PHOTO ACKNOWLEDGMENTS

Image credits: NASA, pp. 4, 6, 18, 19; U.S. Air Force photo by Airman 1st Class Mike Meares/Wikimedia Commons, p. 5; U.S. Space Force photo by Staff Sgt. Alexandra M. Longfellow/United States Department of Defense, p. 7; AP Photo/NewsBase, p. 8; U.S. Army/flickr (CC BY 2.0), p. 9; U.S. Space Force photo by Senior Airman Hanah Abercrombie/United States Department of Defense, p. 10; U.S. Air Force photo by Staff Sgt. Chandler Baker/United States Department of Defense, p. 11; NASA/Lori Losey/flickr (CC BY-SA 2.0), pp. 12–13; REUTERS/Larry Downing/Alamy Stock Photo, p. 14; U.S. Space Force photo by Michael Peterson/United States Department of Defense, p. 15; U.S. Navy photo by Mass Communication Specialist 1st Class Kegan E. Kay/United States Department of Defense, p. 16; U.S. Navy photo by Mass Communication Specialist 1st Class Jen S. Martinez/United States Department of Defense, p. 17; Dave Young/Wikimedia Commons (CC BY 2.0), p. 20. Design elements: anyababii/Getty Images; Andrey_Kuzmin/Shutterstock.com; Andrey_Kuzmin/Shutterstock.com.

Cover: U.S. Navy photo by Mass Communication Specialist 1st Class Kegan E. Kay/United States Department of Defense.